Illustrated Poultry Primer

by US Dept. of Agriculture

with an introduction by Jackson Chambers

This work contains material that was originally published in 1919.

This publication is within the Public Domain.

This edition is reprinted for educational purposes and in accordance with all applicable Federal Laws.

Introduction Copyright 2017 by Jackson Chambers

COVER CREDITS

Front Cover
Silver-Laced Wyandotte Rooster by ripperda (wyandotte haan)
[CC BY 2.0 (http://creativecommons.org/licenses/by/2.0)],
via Wikimedia Commons

Back Cover
Golden-Laced Wyandotte by Inohae (Self-photographed)
(Cropped version used)
[CC BY 4.0 (http://creativecommons.org/licenses/by/4.0)],
via Wikimedia Commons

Research / Sources
Wikimedia Commons
www.Commons.Wikimedia.org

Many thanks to all the incredible photographers, artists,
researchers, and archivists who share their great work.

PLEASE NOTE :
As with all reprinted books of this age that are intended to perfectly reproduce the original edition, considerable pains and effort had to be undertaken to correct fading and sometimes outright damage to existing proofs of this title. At times, this task can be quite monumental, requiring an almost total rebuilding of some pages from digital proofs of multiple copies. Despite this, imperfections still sometimes exist in the final proof and may detract slightly from the visual appearance of the text.

DISCLAIMER :
Due to the age of this book, some methods or practices may have been deemed unsafe or unacceptable in the interim years. In utilizing the information herein, you do so at your own risk. We republish antiquarian books without judgment or revisionism, solely for their historical and cultural importance, and for educational purposes.

Self Reliance Books

Get more historic titles on animal and stock breeding, gardening and old fashioned skills by visiting us at:

http://selfreliancebooks.blogspot.com/

INTRODUCTION

I am so pleased to present to you another reprint of an important poultry publication – *Illustrated Poultry Primer*. It was released in 1919 by the U.S. Department of Agriculture, and is just shy of a century old. It is also known as *Farmers' Bulletin 1040*.

More and more people are turning to raising their own poultry these days, so that they know just what is in the food they are feeding their family. There is currently a renaissance in back-yard farming, with folks raising their own chickens for meat and eggs, their own goats for milk and cheese, their own cattle for meat, or growing their own fruits and vegetables. If you have enough space, you can do all four!

Features chapters on *Selecting the Breed, Breeding, Artificial and Natural Incubation and Breeding, Poultry Houses and Fixtures, Feeding for Egg Production, Caponizing, Common Diseases and Treatment,* and more.

Always informative, the this USDA publication is a great primer for beginners in the poultry business, or for anybody considering raising their own poultry.

Jackson Chambers,
State of Jefferson, November 2017

THE OBJECT of this bulletin is to give, by means of photographs and brief statements, the fundamentals underlying the production of poultry.

An effort has been made to illustrate the various phases of poultry production in such a way as to impress upon the reader's mind the principles of poultry keeping.

Under "Selecting the Breed," for example, photographs are shown of the more popular breeds of each of the three main classes of poultry, giving the reader an immediate and complete idea of the appearance of these fowls, the classes to which they belong, and their economical usefulness. In like manner other essential phases of poultry keeping are illustrated and discussed.

Throughout the bulletin references are given to other publications issued by this department which give more detailed information on each of the subjects discussed and which may be obtained on request.

Contribution from the Bureau of Animal Industry
JOHN R. MOHLER, Chief

Washington, D. C.   March, 1919

ILLUSTRATED POULTRY PRIMER

Harry M. Lamon and Jos. Wm. Kinghorne,

Animal Husbandry Division.

CONTENTS.

	Page.		Page.
Selecting the breed	3	Feeding for egg production	19
Breeding	8	Marketing the product	22
Artificial and natural incubation and brooding	10	Caponizing	24
		Lice and mites	25
Poultry houses and fixtures	14	Common diseases and treatment	26
Produce infertile eggs	16	Nine essential features	28

SELECTING THE BREED.

In the selection of a breed or variety of poultry care should be taken to obtain healthy, vigorous stock.

Beginners are urged to keep but one variety of a breed of fowls. There is no best breed of poultry. Select the breed that suits your purpose best.

Be sure that the male bird at the head of the flock is standard bred.

Mongrel male.

Standard-bred male.

A standard-bred male at the head of a mongrel flock will improve the quality of the stock materially. A mongrel male will produce no improvement in quality.

Given the same care and feed, standard-bred fowls will make a greater profit than mongrel fowls.

A standard-bred flock (upper); a mixed or mongrel flock (lower).

STANDARD-BRED fowls produce uniform products which bring higher prices.

Standard-bred stock and eggs, sold for breeding purposes, bring higher prices than market quotations.

Standard-bred fowls can be exhibited and thus compete for prizes.

THE products from mongrel fowls are not uniform and do not always bring the highest prices.

Eggs and stock from mongrel fowls are not sold for breeding purposes

Mongrel fowls are not exhibited in poultry shows or exhibits.

THE GENERAL-PURPOSE BREEDS.

The general-purpose breeds are best suited to most farms where the production of both eggs and meat is desired. The four most popular representatives of this class are the Plymouth Rock, Wyandotte, Orpington and Rhode Island Red.

Plymouth Rock.

Wyandotte.

Orpington.

Rhode Island Red.

All these breeds, with the exception of the Orpington, are of American origin. They are characterized by having yellow skin and legs, and lay brown-shelled eggs. The Orpington is of English origin, has a white skin, and also lays brown-shelled eggs.

For detailed discussion of the various breeds of fowls of American origin request Farmers' Bulletin 806 on "Standard Varieties of Chickens. I. The American Class," which may be had on application to the U. S. Department of Agriculture, Washington, D. C.

THE EGG BREEDS.

The Mediterranean or egg breeds are best suited for the production of white-shelled eggs. Representatives of this class are bred largely for the production of eggs rather than for meat production. Among the popular breeds of this class are: Leghorn, Minorca, Ancona, and Andalusian.

Leghorn.

Minorca.

Ancona.

Andalusian.

One of the outstanding characteristics of the egg breeds is the fact that they are classed as nonsitters; that is, as a rule they do not become broody and hatch their eggs. When fowls of this class are kept, artificial incubation and brooding are usually employed.

For detailed discussion of the various breeds of this class request Farmers' Bulletin 898 on "Standard Varieties of Chickens. II. The Mediterranean Class," which may be obtained on application to the U. S. Department of Agriculture, Washington, D. C.

THE MEAT BREEDS.

The meat breeds of poultry are primarily kept for the production of meat rather than for the production of eggs in large quantities. Representatives of this class are: Langshan, Brahma, Cochin, and Cornish.

Langshan.

Brahma.

Cochin.

Cornish.

Although classed as meat breeds representatives of this class are sometimes kept as general-purpose fowls. Each of these breeds is heavier and larger in size than the egg breeds or those of the general-purpose class, and lay brown-shelled eggs.

For further information on the various breeds of this class, request Farmers' Bulletin on "Standard Varieties of Chickens. III. The Asiatic, English, and French Classes," which may be obtained on application to the U. S. Department of Agriculture, Washington. D. C.

BREEDING.

Fowls for breeding purposes should be strong, healthy, vigorous birds. The comb, face, and wattles should be of a bright-red color, eyes bright and fairly prominent, head comparatively broad and short and not long or crow-shaped, legs set well apart and straight, plumage clean and smooth.

Females showing high and low vitality. The latter to be avoided when selecting females for breeding.

A knock-kneed fowl. The kind to be avoided as a breeder.

Defects of the kind shown here should be avoided in selecting breeders.

If possible, free range should be provided for the breeding pen.

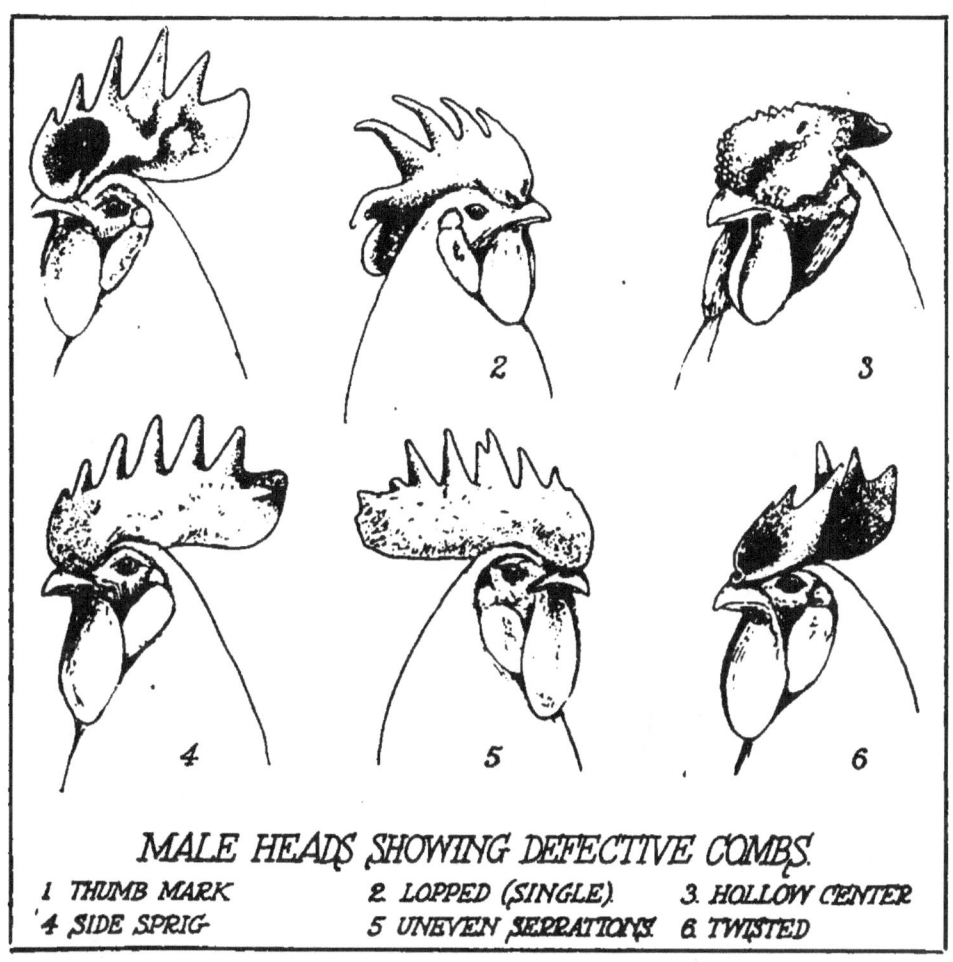

MALE HEADS SHOWING DEFECTIVE COMBS.
1 THUMB MARK 2 LOPPED (SINGLE) 3 HOLLOW CENTER
4 SIDE SPRIG 5 UNEVEN SERRATIONS 6 TWISTED

Usually hens make better breeders than pullets. Cockerels, if well grown and matured, often give better fertility than older birds. However, cock birds that have proved good breeders should be used.

When the breeding flock is confined to a yard, the size of the mating should be 1 male to 10 or 12 females. When allowed free range, the number of females can be increased to 20 or 25 with good results.

Matings should be made two weeks before the eggs are saved for hatching.

MALES WITH DEFECTIVE TAIL CARRIAGE
1 SQUIRREL 2 WRY

ARTIFICIAL AND NATURAL INCUBATION AND BROODING.

Have everything ready beforehand and start your hatching operations early in the year. In sections where the climate is temperate, February, March, and April are the best months for hatching. The early hatched pullet is the one that begins to lay early in the fall and continues to lay when eggs are high in price.

Select uniform, fairly large sized eggs for hatching.

Operate the incubator according to the manufacturer's directions to produce the best results.

A well-ventilated cellar of uniform temperature is an excellent place to operate the incubator.

Test the eggs for fertility on the seventh and fourteenth days.

Do not open the incubator after the eighteenth day until the chicks are hatched.

Given proper care and attention, the hen is the most valuable incubator for the farmer whose poultry operations are of moderate size.

In cool weather place from 10 to 13 eggs under the hen; in warm weather from 13 to 15 eggs.

Chicks should not receive feed until they are 36 hours old.

When artificial incubation is used, start the brooder a day or two before putting in the chicks, to see that the heating apparatus is working properly. Brooder lamps should be cleaned every day.

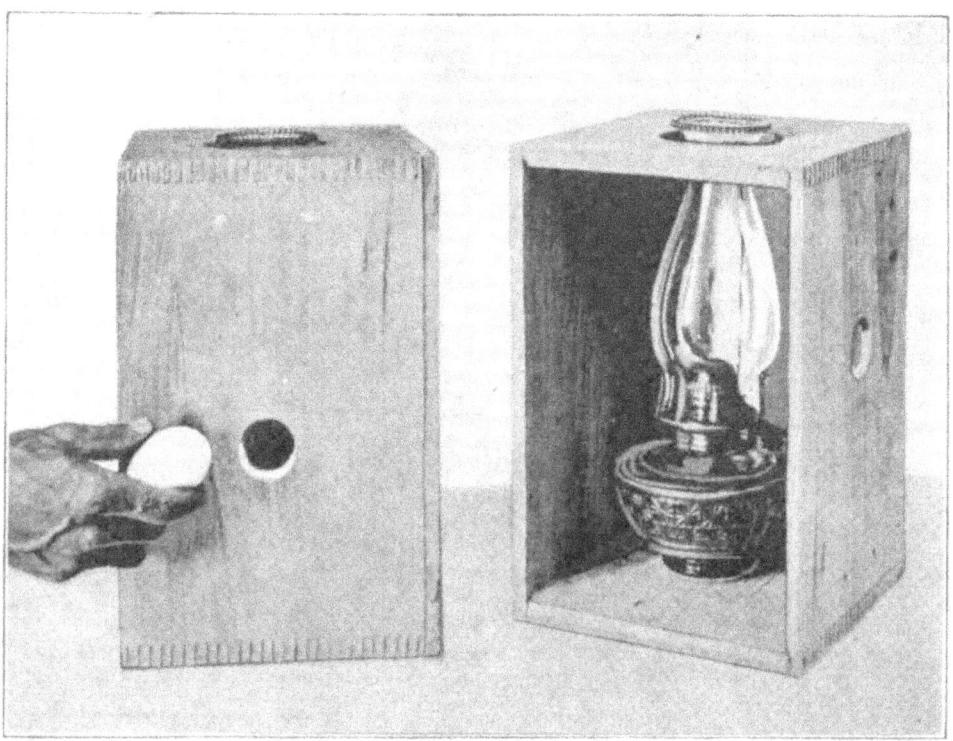
Homemade egg candler. The hole for testing eggs should be directly opposite the flame of the lamp.

A good hatch.

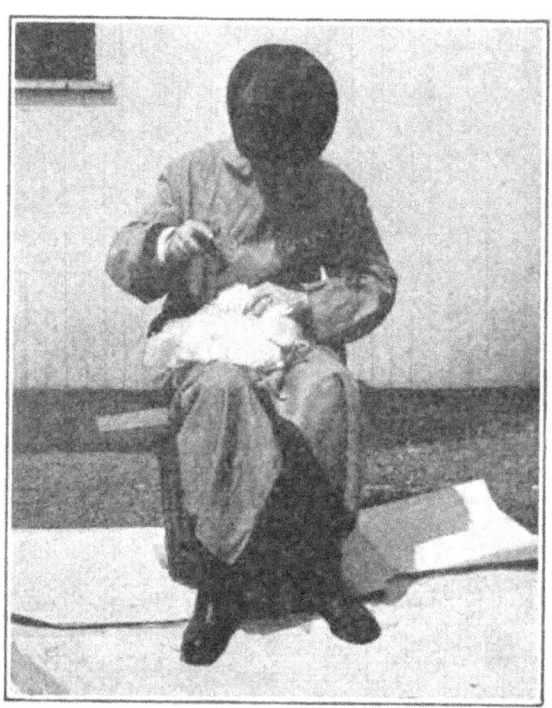

Dust the hen thoroughly with a good lice powder before placing her on the nest.

An excellent range providing shade and shelter for growing chicks.

If possible locate the brooders on ground that has recently been cultivated, thereby eliminating the danger of tainted soil and possible disease.

Chicks having access to a shaded range, such as shown above, develop and thrive better in warm weather than those not having such range.

Do not allow the mother hen to range over the farm with the chicks.

Confine the mother hen to a brood coop until the chicks are weaned.

In the case of hen-hatched broods, the coop for hen and chicks should be well ventilated, easy to clean, and large enough to insure comfort. To allow the hen to range over the farm with the chicks will often be the cause of heavy losses.

For the first three days chicks may be fed a mixture of equal parts of hard-boiled eggs and rolled oats or stale bread, or stale bread soaked in milk. When bread and milk are used, care should be taken to squeeze all the milk out of the bread. From the third or fourth day commercial chick feed may be fed until the chicks are old enough to eat wheat screenings or cracked corn.

To insure rapid and uniform growth of the chicks, provide in addition to a grain feed a dry mash to which the chickens will have access at all times.

For additional information on incubation and brooding, request Farmers' Bulletins 585 and 624.

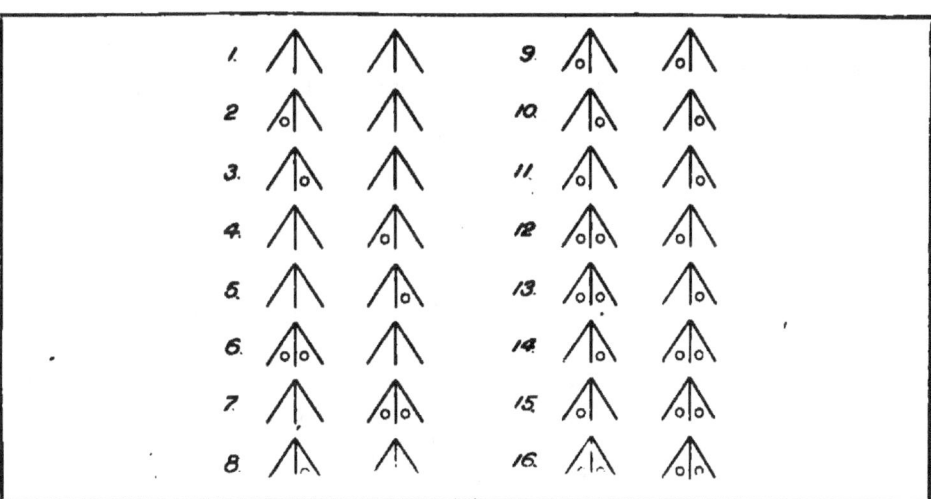

Toe-mark the chicks as soon as they are hatched. This enables one to tell their ages later.

POULTRY HOUSES AND FIXTURES.

Select a location for the poultry house that has natural drainage away from the building. A dry, porous soil, such as sand or gravelly loam, is preferable to a clay soil.

Rebuilding a poultry house out of old lumber at small cost.

The building should face the south or southeast to insure the greatest amount of sunlight during the winter.

The roosts should be built on the same level, about 3 feet from the floor, with a droppings board about 6 inches below the roosts.

A good interior arrangement for a poultry house, showing roosts and droppings boards with nests underneath and wire coop at end for confining broody hens. Note ventilators in back of house and the abundance of sunlight, which insures a dry house and healthy fowls.

Good roosts may be made of 2 by 2 inch material with the upper edges rounded.

The nests may be placed on the side walls or under the droppings boards. It is best to have them darkened, as hens prefer a secluded place in which to lay. For further information on poultry house construction request Farmers' Bulletin 574.

A partial open-front curtain house is conceded to be the best type for most sections of temperate climate.

PRODUCE INFERTILE EGGS!

FERTILE EGGS SPOIL QUICKLY IN SUMMER WEATHER.

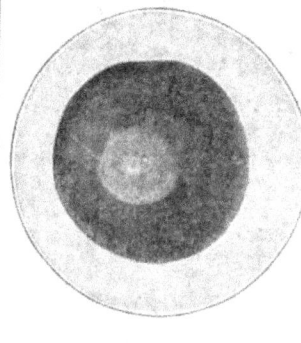

No. 1. Fertile egg after 24 hours at 103° F. Fertile germ beginning to hatch. Not perfect for food.

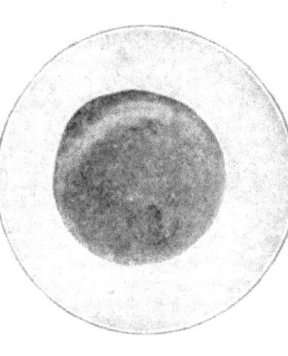

No. 2 Fertile egg after 36 hours at 103° F. Blood ring formed. Not good for food.

INFERTILE EGGS KEEP BETTER THAN FERTILE EGGS IN SUMMER.

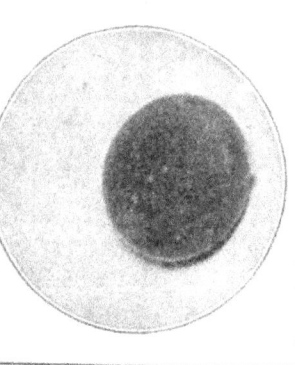

1A. Infertile egg after 24 hours at 103° F. No fertile germ. No blood ring. Still good food. It would be still better if kept cool.

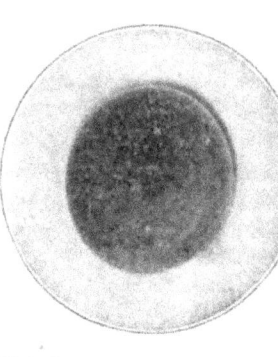

2A. Infertile egg after 36 hours at 103° F. Compare with fertile egg under the same conditions.

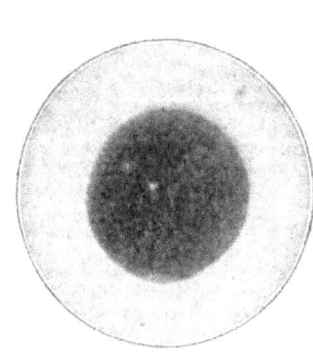

FRESH EGG.

FERTILE EGGS COST THE FARMER $15,000,000 A YEAR.

Farmers lose $45,000,000 annually from bad methods of producing and handling eggs. One-third of this loss is preventable, because it is due to the partial hatching of fertile eggs which have been allowed to become warm enough to begin to incubate.

The rooster makes the egg fertile.

season is over.

The rooster does not help the hens to lay. He merely fertilizes the germ of the egg. The fertile germ in hot weather quickly becomes a blood ring, which spoils the egg for food and market. Summer heat has the same effect on fertile eggs as the hen or incubator.

INFERTILE EGGS WILL NOT BECOME BLOOD RINGS.

After the hatching season, cook, sell, or pen your rooster. Your hens not running with a male bird will produce infertile eggs—quality eggs that keep best and market best.

RULES FOR HANDLING EGGS ON THE FARM.

Heat is the great enemy of eggs, both fertile and infertile. Farmers are urged to follow these simple rules, which cost nothing but time and thought and will add dollars to the poultry yard returns:

1. Keep the nests clean; provide one nest for every four hens.
2. Gather the eggs twice daily.
3. Keep the eggs in a cool, dry room or cellar.
4. Market the eggs at least twice a week.
5. Sell, kill, or confine all male birds as soon as the hatching season is over.

NOTICE.

Valuable published information on the raising and care of poultry and eggs and individual advice on these subjects may be obtained by writing to the

Animal Husbandry Division,
UNITED STATES DEPARTMENT OF AGRICULTURE,
Washington, D. C.

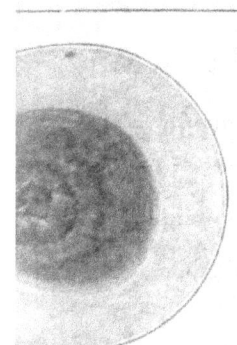

No. 3. Fertile egg after 48 hours at 103° F. Blood ring fully developed. Unfit for market. Will be thrown out by candler.

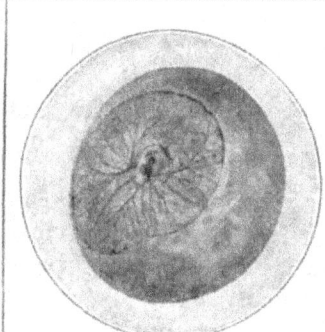

No. 4. Fertile egg after 72 hours at 103° F. Blood vessels of embryo chick clearly marked.

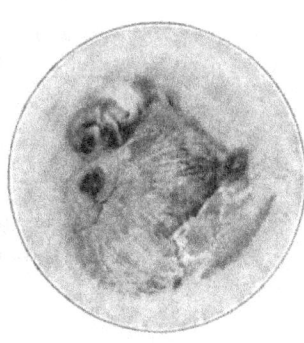

No. 5. Fertile egg after 7 days at 103° F. Compare with infertile egg and fresh egg.

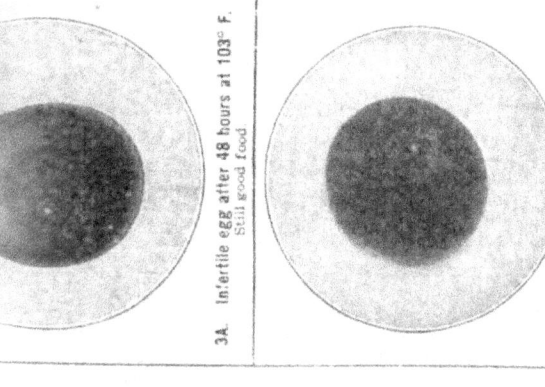

3A. Infertile egg after 48 hours at 103° F. Still good food.

4A. Infertile egg after 72 hours at 103° F. Not as absolutely fresh egg, but useful in cookery.

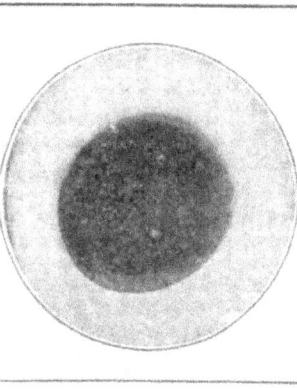

5A. Infertile egg after 7 days at 103° F. Still usable for food. It would be a perfect egg if it had been kept cool. Compare with fresh egg and fertile egg.

TRAP NESTS.

A trap nest is a laying nest so arranged that after a hen enters it she is confined until released by the attendant. The trap nest shown in the accompanying illustration is used with good results on the Government poultry farm and is very similar to the nest used at the Connecticut State experiment station. It is very simple and may be built at a small cost.

Trap nests enable the poultryman to distinguish between the layers and the drones.

When possible it is advisable to trap-nest the layers for the following reasons:

1. To tame the birds, thereby tending toward increased egg production.
2. To furnish definite knowledge concerning traits and habits of individuals.
3. To furnish the only satisfactory basis for utility or other breeding.
4. To eliminate the nonproductive hen.
5. To add mechanical precision to judgment and experience in developing and maintaining the utility of a flock.

For further information and plans showing the construction of a trap nest, send for Farmers' Bulletin 682, "A Simple Trap Nest for Poultry."

FEEDING FOR EGG PRODUCTION.

CLASSIFICATION OF POULTRY FEEDS.

Nature provides—	Scientific classification.	Poultrymen feed—
Worms and bugs..	Nitrogenous material, or protein....	Meat (green cut bone or beef scrap), milk or cottage cheese.
Seeds	Carbohydrates	Wheat, oats, corn, barley, etc.
Greens	Succulents	Lettuce, cabbage, kale, mangels, alfalfa, clover, sprouted oats, etc.
Grit	Mineral matter	Grit and oyster shell.
Water	Water	Water.

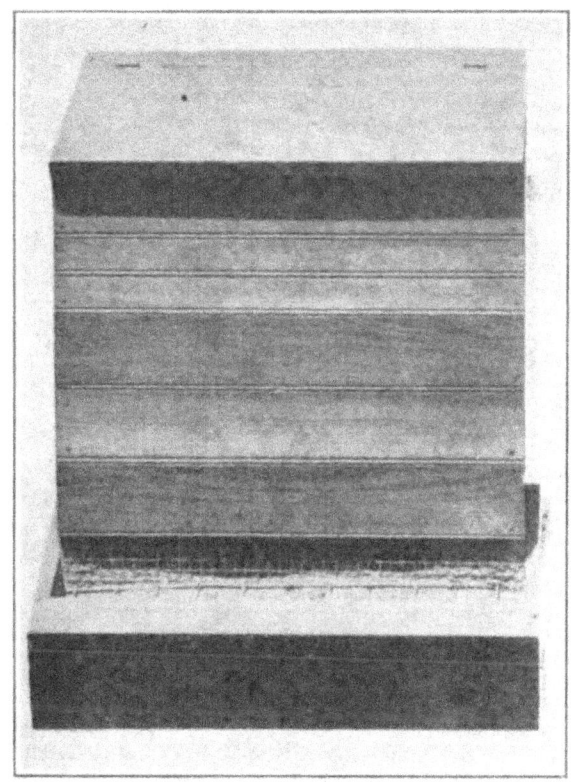

A homemade dry-mash hopper.

Oats in the process of sprouting.

In order to obtain an abundance of eggs it is necessary to have healthy, vigorous stock, properly fed.

The following are good grain mixtures for the laying stock, the proportions being by weight:

Ration 1.	Ration 2.	Ration 3.
Equal parts of:	3 parts cracked corn.	2 parts cracked corn.
Cracked corn.	2 parts oats.	1 part oats.
Wheat.	1 part wheat.	
Oats.		

A choice of any one of these rations should be scattered in the litter twice daily, morning and evening.

Average amount of feed consumed by a laying hen and eggs produced.

Either of the following suggested dry-mash mixtures should be fed in a dry-mash hopper such as illustrated, allowing the fowls to have access to it at all times.

Mash No. 1.	Mash No. 2.
2 parts corn meal.	3 parts corn meal.
1 part bran.	1 part beef scrap.
1 part middlings.	
1 part beef scrap.	

When fowls do not have access to natural green feed, sprouted oats, cabbage, mangels, cut clover, etc., should be fed.

When wet mashes are fed, be sure that they are crumbly and not sticky. Plenty of exercise increases the egg yield.

Fresh, clean drinking water should be always provided. Charcoal, grit, and oyster shell should be placed before the fowls so that they can have access to them at all times.

For additional information on feeds and feeding request Farmers' Bulletin 287, "Poultry Management," and Farmers' Bulletin 528, "Hints to Poultry Raisers," from U. S. Department of Agriculture, Washington, D. C.

A rural cafeteria.

Relative Losses of Fertile Compared With Infertile Eggs.

Produce the infertile egg. Infertile eggs are produced by hens that have no male birds with them. See pages 16 and 17.

The following table shows that the losses of fertile eggs are computed to be nearly twice as great as in the case of infertile eggs.

	Fertile eggs.	Infertile eggs.
	Per cent	Per cent
On the farm	29.0	15.5
At country store	7.1	4.0
Transportation to packing house	6.4	4.7
Total	42.5	24.2

To produce infertile eggs confine or dispose of the male birds. This has no influence on the number of eggs laid by the hens.

MARKETING THE PRODUCT.

The hen's greatest egg-producing periods are the first, second, and third years, depending upon the breed. The heavier breeds, such as Plymouth Rocks, may be profitably kept for two years; the lighter breeds, such as Leghorns, three years.

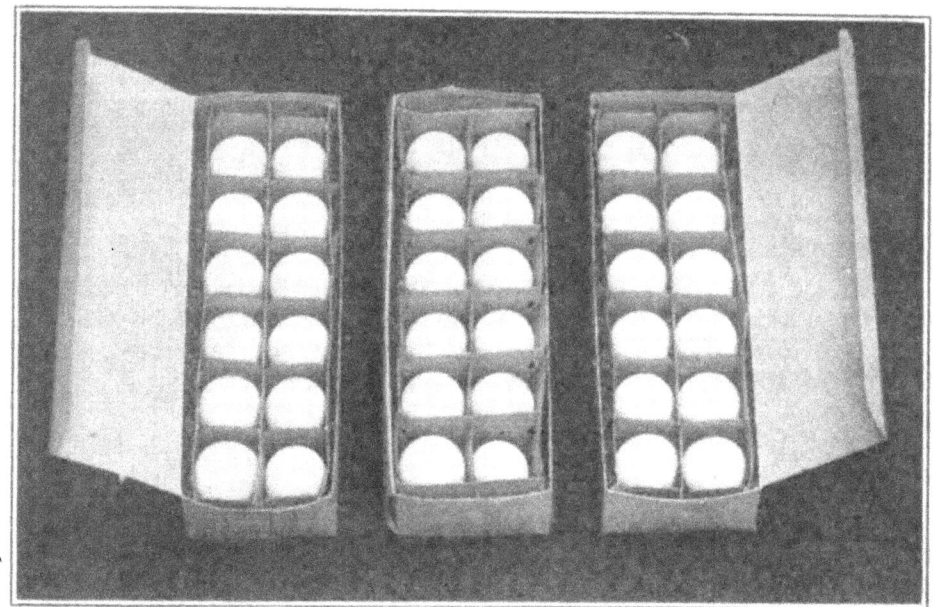

Uniform products command the best prices. Standard-bred fowls produce uniform products.

Market white-shelled and brown-shelled eggs in separate packages. Eggs irregular in shape, those which are unusually long or thin-shelled, or which have shells otherwise defective, should be kept by the producer for home use, so that breakage in transit may be reduced as much as possible.

For additional information on packing and shipping eggs by parcel post request Farmers' Bulletin 830, "Marketing Eggs by Parcel Post," issued by the U. S. Department of Agriculture, Washington, D. C.

Extremely large, small, and soiled eggs should not be marketed; use them at home. All the eggs above were produced by a farm flock of mixed or mongrel fowls.

The result of a trip under the corncrib.

Everybody in the shade except the eggs.

Eggs from "stolen" nests should not be marketed; they are of unknown age and quality and should be used at home.

When taking eggs to market, protect them from the sun's rays in warm weather. Ship or deliver eggs twice or three times weekly.

Notice the candler has places for the good eggs as well as for checks (cracked eggs), dirty eggs and "rots." When selling eggs insist that they be bought on a quality basis.

Infertile eggs will withstand marketing conditions much better than fertile eggs.

All cockerels not intended to be kept or sold for breeders should be marketed when they reach suitable size. Such birds confined in a homemade fattening battery or coop and fed a fattening ration for a week or ten days will not only increase in weight but bring a better price on the market, because of improved quality.

A shipment of eggs on the railroad-station platform, exposed to the sun.

Candling eggs for quality.

CAPONIZING.

A capon is an unsexed male bird, which when mature is of larger size and more desirable for eating than cockerels or cocks.

A Buff Orpington cock.

A Buff Orpington capon.

Boys caponizing a cockerel.

By following directions and with a little practice, poultrymen will find caponizing a simple operation. For detailed information on caponizing, request Farmers' Bulletin 849.

LICE AND MITES.

The free use of an effective lice powder is always advisable. A dust bath, consisting of road dust and wood ashes, is essential in ridding fowls of lice.

Sodium fluorid, a white powder which can be obtained from druggists, is also effective. Apply a pinch of the powder at the base of the feathers on the head, neck, back, breast, below the vent, base of tail, both thighs, and on the underside of each wing.

Applying sodium fluorid.

An effective remedy for lice on chicks is a small quantity of melted lard rubbed under the wings and on top of the chick's head.

The free use of kerosene or crude oil on the roosts and in the cracks of the house will help to exterminate mites.

Whitewash is effective against all vermin.

It is possible and thoroughly practicable to keep the poultry flock reasonably free from lice and mites. Such practices should be the aim of every one who is endeavoring to establish a successful flock of poultry.

For complete information on mites and lice, request Farmers' Bulletin 801.

COMMON DISEASES AND TREATMENT.

All diseased birds should be isolated.

Colds and roup.—Disinfect the drinking water as follows: To each gallon of water add one tablespoonful of sodium sulphite or as much potassium permanganate as will remain on the surface of a dime.

A bad case of roup.

Chicken pox.—Put a touch of iodin on each sore and apply carbolated vaseline.

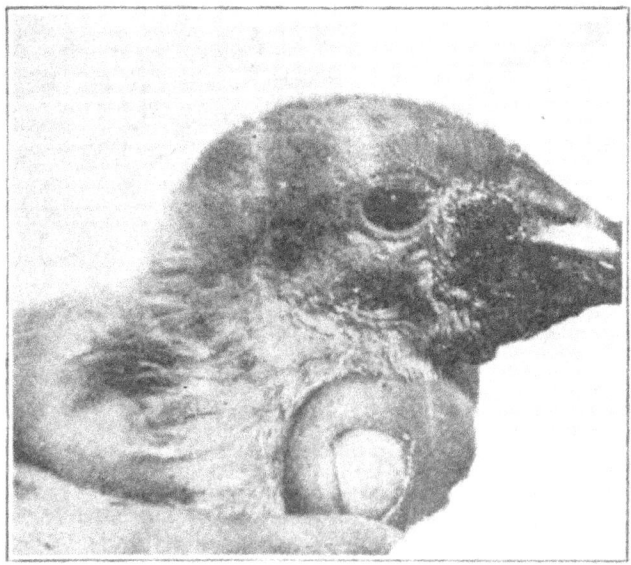

Chicken pox.

Gapes.—Fresh ground and vigorous cultivation will often remedy this trouble, which is caused by small gapeworms that live in the soil and attach themselves to the inside of the throat.

Diarrhea in hens.—Low-grade wheat flour or middlings is good for this trouble. A teaspoonful of castor oil containing 5 drops of oil of turpentine to each fowl is also good.

Bumblefoot.—When the feet are badly swollen, a small cut should be made with a clean, sharp knife, and the pus removed. Wash the wound out with equal parts of hydrogen peroxide and water, grease with vaseline, and bandage.

Limberneck.

Limberneck.—A teaspoonful of caster oil given to the fowl will sometimes effect a cure.

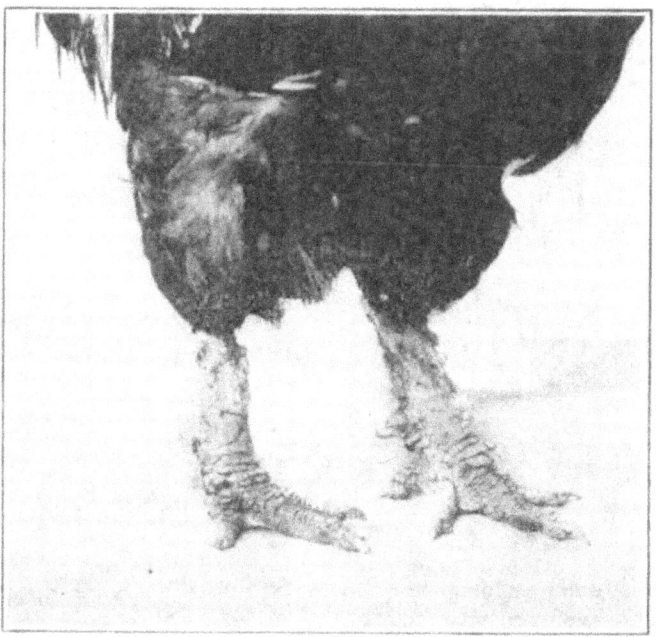
Scaly legs.

Scaly legs.—Apply vaseline containing 2 per cent of creolin to the affected parts and after 24 hours soak in warm, soapy water. Repeat treatment until cured.

For a detailed discussion of the foregoing and other poultry diseases, request Farmers' Bulletin 957, "Important Poultry Diseases."

NINE ESSENTIAL FEATURES FOR PROFITABLE POULTRY KEEPING.

1—KEEP BETTER POULTRY:
   Standard-bred poultry increases production and improves the quality.

2—SELECT VIGOROUS BREEDERS:
   Healthy, vigorous breeders produce strong chicks.

3—HATCH THE CHICKS EARLY:
   Early hatched pullets produce fall and winter eggs.

4—PRESERVE EGGS FOR HOME USE:
   Preserve when cheap for use when high in price.

5—PRODUCE INFERTILE EGGS:
   They keep better. Fertile eggs are necessary for hatching only.

6—CULL THE FLOCKS:
   Eliminate unprofitable producers and reduce the feed bill.

7—KEEP A BACK-YARD FLOCK:
   A small flock in the back yard will supply the family table.

8—GROW YOUR POULTRY FEED:
   Home-grown feed insures an available and economical supply.

9—EAT MORE POULTRY AND EGGS:
   Poultry and eggs are highly nutritious foods.

For further information or individual advice on poultry raising write to your State Agricultural College, or to the Animal Husbandry Division, Bureau of Animal Industry.

UNITED STATES DEPARTMENT OF AGRICULTURE
WASHINGTON, D. C.

www.ingramcontent.com/pod-product-compliance
Lightning Source LLC
Chambersburg PA
CBHW062207220526
45470CB00009B/2959